Loud Animals

Dona Herweck Rice

Publishing Credits

Rachelle Cracchiolo, M.S.Ed., *Publisher*
Conni Medina, M.A.Ed., *Managing Editor*
Nika Fabienke, Ed.D., *Content Director*
Véronique Bos, *Creative Director*
Shaun N. Bernadou, *Art Director*
Seth Rogers, *Editor*
John Leach, *Assistant Editor*
Courtney Roberson, *Senior Graphic Designer*

Image Credits: All images from iStock and/or Shutterstock.

Library of Congress Cataloging-in-Publication Data

Names: Rice, Dona, author.
Title: Loud animals / Dona Herweck Rice.
Description: Huntington Beach, CA : Teacher Created Materials, [2019] | Audience: K to grade 3. |
Identifiers: LCCN 2018029715 (print) | LCCN 2018029828 (ebook) | ISBN 9781493899548 | ISBN 9781493898800
Subjects: LCSH: Animal communication--Juvenile literature. | Animal sounds--Juvenile literature. | Vocabulary.
Classification: LCC QL751.5 (ebook) | LCC QL751.5 .R5475 2019 (print) | DDC 591.59/4--dc23
LC record available at https://lccn.loc.gov/2018029715

Teacher Created Materials
5301 Oceanus Drive
Huntington Beach, CA 92649-1030
www.tcmpub.com

ISBN 978-1-4938-9880-0

Printed in China
Nordica.082018.CA21800936

 call out in

Coyotes

the  .

desert

They call out to

each other.

Dolphins

call out in the

ocean

.

They call out to

each other.

 call out in

Goats

the  .

mountains

They call out to

each other.

 call out in

Squirrels

the  .

forest

They call out to

each other.

 call out in

Parrots

the  .

jungle

They call out to

each other.

High-Frequency Words

New Words

call	each
other	out

Review Words

in	the
they	to